SOTHIS PRESS

Sothis Press aims to completely reimagine the public domain book with an emphasis on universal access to captivatingly designed print editions. We believe that technology will enable every single book ever published to be perpetually accessible in beautiful print form. Sothis is beauty in ubiquity.

www.sothis.press
1 Waterhouse Square, London, ECIN 2ST

Pocket Handbook of Blowpipe Analysis

Designed for the Use of Students and Prospectors With the Idea of Making Oral Instruction Unnecessary

G. Montague Butler

SOTHIS PRESS

ISBN 978-1-0368-0010-9

Pocket Handbook of Blowpipe Analysis
is in the public domain.

This edition is published by
Sothis Press © 2023. The company
is registered in England and Wales.
The Company Number is 14900389.

POCKET HANDBOOK
OF
BLOWPIPE ANALYSIS

DESIGNED FOR THE USE OF

STUDENTS AND PROSPECTORS WITH THE IDEA OF MAKING ORAL INSTRUCTION UNNECESSARY

BY

G. MONTAGUE BUTLER, E.M.
Dean, College of Mines and Engineering, University of Arizona

FIRST EDITION, CORRECTED
SECOND THOUSAND

NEW YORK
JOHN WILEY & SONS, INC.
LONDON: CHAPMAN & HALL, LIMITED
1916

PREFACE

THIS little book was written, primarily, to satisfy the demands of those instructors who have been using the author's Pocket Handbook of Minerals as a text-book for courses in mineralogy. While there is no doubt that a thorough knowledge of the *physical* characteristics of minerals should be the end sought by all teachers and students of this subject, it is often desirable to be able to fall back upon other simple tests in corroboration of conclusions reached by observation, or when studying an unfamiliar mineral. For this purpose, blowpipe analysis is very satisfactory, since the necessary implements and reagents are comparatively few and simple and may be so selected as to be portable.

While there are many works on this subject, they are either too comprehensive for the purpose for which this is intended, or else form merely an introductory chapter to a work on mineralogy. In both cases, their purchase entails a needlessly high expense, and their directions and statements are often so vague and incomplete as to require continual explanation by an instructor.

No originality is claimed for the tests themselves as given in this pamphlet; they are the same as are included in all books on the subject, and as have been taught in the Colorado School of Mines for years by Professor H. B. Patton. Such modifications and additions have been made, however, as experience has proven desirable. The text and plan of the work are, of course, original, and so are most of the data included in the notes on the various tests. Utility and conciseness have been the ends sought, and it is believed that no superfluous details have been included and that nothing essential has been omitted.

Secondarily, the book was written to satisfy the needs of miners and prospectors, whose first question on finding a new mineral is, "What does it contain?" It is so elementary in its nature and the directions are so complete that anyone with a common-school education, the proper instruments, and this book should be able in most cases to answer this question for himself. It was for this class of readers alone that Chapters V and VI were included.

Finally, it is believed that assayers and chemists will find the book useful in making preliminary examinations of unknown substances.

TABLE OF CONTENTS

BLOWPIPE ANALYSIS

CHAPTER I

BLOWPIPE INSTRUMENTS, REAGENTS, AND OPERATIONS

NEARLY all dealers in assayers' or chemists' supplies carry sets and separate pieces of blowpipe apparatus, and many pieces may be obtained elsewhere, as will be seen from the following brief description of the articles used in the operations later described. Most of the sets now on the market are either too elaborate, too bulky, or else are impractical, and great care should be exercised in their selection.

Blowpipe. Many types are manufactured, and each may have its own peculiar advantages, but almost any one in which the aperture is not too large or too small will answer the purpose. The most satisfactory type has a trumpet-shaped mouthpiece, a small chamber in which the saliva may accumulate, and a removable tip. In portable sets it is not practicable to provide a blowpipe with a trumpet-shaped mouthpiece, however, and this may be dispensed with, although somewhat easier on the lip muscles. Some makes are provided with platinum tips, but this is an unnecessary refinement unless a great deal of work is to be done.

A piece of very fine platinum or steel wire is the best implement with which to clean out a clogged tip.

Lamps. It is desirable, although not absolutely essential, to have two lamps, one for oil and the other for alcohol. The latter is handier than the oil lamp for a few operations which will be mentioned later, but in every case it can be replaced by the oil lamp without material disadvantage.

The oil lamp should have a rectangular wick opening about half an inch long and less than half as wide; the wick opening in the alcohol lamp may be of any shape. Care should be taken (particularly with the oil lamp) not to have the wick so tight as to impede the flow of oil, to trim off charred wick or irregularities as fast as they form, and to keep the wick just high enough not to smoke. The best fuel for the oil lamp is a mixture composed of two parts of lard oil and one part of kerosene. This solidifies in very cold weather, but thaws out soon after lighting the lamp.

Any other flame, such as that from a candle or kerosene lamp, may be used instead of these lamps, and it is often necessary to resort to these in the field, but the lamps and oil described will give the best results in the laboratory. In some places gas blowpipes are used, but one who has attained proficiency with such apparatus is lost in the field, so simpler instruments are preferable.

Platinum-tipped Forceps. The most convenient type of these has tips of platinum on one end and of base metal on the other, the platinum-tipped end being provided with a spring which holds the tips together. Precautions as to the use of these forceps are given later. They soon become discolored with use, but, if the precautions just mentioned are observed, this will not harm them. They may be cleaned by using very fine sandpaper or scraping with a knife-blade.

Platinum Wire and Holder. The wire should be of about 26 American or B. & S. Wire Gauge in thickness, and should be cut into pieces between two and three inches long. The holder may be a glass tube into which one end of the wire is fused, but a mechanical holder with a hollow handle in which extra wires may be kept is more convenient. Precautions concerning the use of the platinum wire are given later.

Charcoal Supports. These should be made from soft wood and should be at least three inches long, preferably more. They should not fissure, break, smoke, or ignite readily in the flame, and should leave little ash when burned.

Unless liquid reagents have been used on the

charcoal in tests resulting successfully, the charcoal may be used many times by scraping off the surface and the deposits formed thereon. Where liquids have been used in successful tests, they are apt to sink for a considerable distance into the charcoal, and to cause a duplication of the test even after a considerable depth of the charcoal has been removed.

Closed Tubes. These are usually formed of three-eighths inch glass tubing three or four inches long, an inch of one end being bent to one side and closed by fusion. Equally satisfactory results are secured from an implement formed by fusing together one end of an open tube (see below). Closed tubes cannot well be cleaned and should be discarded after use. A strip of asbestos, or even paper, wrapped around the upper part of the tube makes a convenient holder for hot tubes.

Open Tubes. These are pieces of three-eighths inch glass tubing three to five inches long. They should be discarded after use unless the results have been negative, when the other end may be used for a new test. The asbestos or paper holder mentioned above should be used for hot tubes.

Miscellaneous. A small slab of hard steel with at least one polished surface for use as an anvil.

A small steel hammer with a flat face.

A small horseshoe or bar magnet. The latter may be procured with one end so shaped as to form a charcoal borer, but this is not necessary.

Test-tubes.

A hand-lens.

A small piece of dark blue glass.

Reagent Bottles and Reagents. Wide-mouth, glass-stoppered bottles are the best in which to keep the dry reagents, which should include powdered (preferably dehydrated) borax, sodium carbonate, sodium ammonium phosphate (salt of phosphorus), acid potassium sulphate, and bismuth flux (equal proportions of potassium iodide and sulphur).

The wet reagents should be kept in glass bottles with glass stoppers, and a dropper stopper will be found a great convenience, although a satisfactory dropper can be easily made from a small glass tube. The following reagents are needed:

Hydrochloric (muriatic) acid. The concentrated acid should be diluted with an equal volume of water for most purposes.

Nitric acid. The concentrated acid is usually employed.

Sulphuric acid (oil of vitriol). For most purposes the concentrated acid should be diluted with four volumes of water. A great deal of heat is generated when water and sulphuric acid are mixed, and this should be done with care. The acid should be added gradually to the water, stirring constantly. Water should *never* be added to sulphuric acid.

Cobalt nitrate. The dry salt should be dissolved in ten parts of water for use.

All acids should be handled with care, as they are more or less corrosive and are capable of inflicting painful injuries when spilled on the skin. When this happens, or if they fall upon fabrics, their effects may be neutralized by moistening with ammonia and then washing thoroughly with water.

Blowpipe Operations. The blowpipe is used for the purpose of concentrating the flame into a long, slender cone which can be readily directed against the substance to be heated. It is very important that the blast be continuous and uniform, although this may seem very difficult at first. The blast is not produced by the lungs, but results from a bellows-like action of the distended cheeks. During the operation, air is inhaled only through the nose, and is exhaled largely through the mouth and the blowpipe. Before trying to use that instrument, distend the cheeks, and, keeping the mouth closed, breathe through the nose for a moment; then open the lips just enough to allow a little air to escape slowly, and admit air from the lungs by a kind of gulping action just fast enough to keep the cheeks fully distended. This may take some practice, but, when it is possible to allow air to escape continuously from the mouth in this way no matter whether it is being exhaled or inhaled through the nostrils, it is time to begin to use the blowpipe.

Producing the Oxidizing Flame. Place the oil lamp so that the longer dimension of the wick is from right to left, and set its right-hand edge upon a pencil or some other low support so that it will tip somewhat to the left. Insert the tip of the blowpipe about one-eighth of an inch within and just above the right-hand side of the wick, and blow steadily parallel to the wick, directing the flame to the left, and producing a clear blue flame about an inch long. If all of the flame cannot be thus diverted to the left, or if there are yellow streaks in the flame, trim or lower the wick. If the whole flame is inclined to be yellow, move the tip of the blowpipe a trifle to the left. If it is impossible to produce a flame approaching the length mentioned above, the opening in the end of the blowpipe is too small, and this opening is too large when a very long, hissing flame is produced. In order to succeed in blowing a steady flame, the hand must rest upon some support, or the third and fourth fingers may be placed against the lamp.

In analytical operations it is sometimes desirable to oxidize substances to be tested, and at other times the aim is to reduce them to the metallic condition; either result can be more or less readily obtained with the blowpipe.

A flame produced in the manner above described is called an oxidizing flame, but the action of all portions of such a flame is not oxidizing. The blue cone contains considerable carbon monoxide and is feebly reducing in its action, but just outside of the blue cone at the tip of the flame is an extremely hot but nearly colorless zone which is strongly oxidizing because of the free oxygen there present, and anything held in this zone about a quarter of an inch from the tip of the blue flame will be in the most favorable position for oxidation.

The oxidizing flame is hotter than the reducing, and the hottest part of this flame is just outside of the blue cone. In the absence of other instructions, substances should always be heated there.

Producing the Reducing Flame. Hold the tip of the blowpipe about one-sixteenth of an inch above and to the right of the wick, and a long, yellow flame containing much unconsumed carbon will be produced. This is sometimes called the smoky reducing flame. Where greater heat is required, the inner cone of the oxidizing flame should be used. The strongest reducing action will take place at the tip of, and within, the yellow cone of the reducing flame.

Other Operations. These will be described in detail when the various tests are discussed.

CHAPTER II

METHODS OF TESTING FOR THE VARIOUS ELEMENTS WITH THE BLOWPIPE

THE methods of testing for the various elements with the blowpipe comprise blowpipe analysis, and many tests are included under this term, including a few in which the blowpipe is not required. The most useful are included in the following list, and will be discussed in the order named:

I. Treatment on charcoal without flux.
II. Treatment on charcoal with flux.
III. Tests in closed tubes.
IV. Tests in open tubes.
V. Tests with borax beads.
VI. Tests with salt of phosphorus beads.
VII. Flame tests.
VIII. Cobalt nitrate coloration tests.
IX. Tests with acids.

The tests given should make it possible to recognize the following elements and substances in most of their combinations:

Aluminum (Al), antimony (Sb), arsenic (As), barium (Ba), bismuth (Bi), boron (B), cadmium (Cd), calcium (Ca), chromium (Cr), cobalt (Co), copper (Cu), flourine (F), gold (Au), iron (Fe), lead (Pb), lithium (Li), magnesium (Mg), manganese (Mn), mercury (Hg), molybdenum (Mo), nickel (Ni), phosphorus (P), potassium (K), selenium (Se), silver (Ag), sodium (Na), strontium (Sr), sulphur (S), tellurium (Te), tin (Sn), titanium (Ti), tungsten (W), uranium (U), vanadium (V), zinc (Zn), and water, silicates, and carbonates.

1. Treatment on Charcoal without Flux

A piece, the size of a BB shot or smaller, of the substance to be tested is pressed into the face of the charcoal about half an inch from one end, or it may be placed in a tiny cavity formed at that point, the object of embedding it somewhat in the charcoal being merely to prevent the particle from sliding off or being blown away. The charcoal is then held in the left hand, pointing right and left, and the blowpipe flame is blown upon the particle, which should be at the right end, in such a manner that the flame is not parallel with the charcoal but impinges downward upon the particle at a small angle.

The piece tested is known as the assay.

Heat the assay strongly for a minute or more in the oxidizing flame, noting any odor or colored flame that may be given off, and then examine the charcoal for any coatings, known as sublimates, that may have been deposited thereon. If no very positive results are thus obtained, repeat the operation, using the reducing flame.

Care must be taken not to place the assay in deep holes that have been burned or scraped into the charcoal. If this is done, the volatilized material will shoot up into the air, and no sublimate will deposit on the charcoal.

Most charcoal forms a little white ash when burned; this should not be confused with a sublimate.

Decrepitation (flying to pieces) of the assay may sometimes be prevented by heating it very slowly, i.e., holding it three or four inches from the flame at first and gradually bringing it nearer. Another method, often satisfactory, consists of blowing the flame against the upper part of the end of the stick of charcoal until it is red hot, thus gradually warming the assay.

If the substance still decrepitates or if no results are obtained by the above methods of treatment, powder the material to be tested *very fine* and press a small amount of the powder onto one end of the charcoal, using a knife blade or spatula for this purpose and forming a flat cake of the powder. Then proceed as before. If the material still decrepitates, moisten it with water and heat very slowly.

Since chlorides of lead, copper, and other substances yield white sublimates on charcoal that may be confused with those mentioned below, it is essential that the charcoal tests be made before any hydrochloric acid has been put on the substance to be tested.

Not only is it necessary to note the color of any sublimates produced, but their volatility should also be tested, as some are very volatile (can be burned off by applying the blowpipe flame for a fraction of a second) while others are quite non-volatile (require the application of considerable heat to entirely remove them) in the oxidizing flame, which should be used for this test.

The results obtained by any of the above means may be thus interpreted:

a. White, very volatile, light sublimate, deposited some distance from the assay. Strong odor of garlic .... As

b. White sublimate with a blue border deposited close to the assay, heavier and less volatile than that yielded by As. Assay will often continue to give forth white fumes after heating has ceased .... Sb

c. White, fairly easily volatilized, heavy sublimate near the assay, shading into a lighter, more volatile coating further out. Odor of garlic ....Sb with As

d. White and yellowish, crystalline (particles are coarse and sharp-edged) sublimate which is changed to a deep ultramarine blue if touched for a *fraction of a second* with the reducing flame. A copper-red coating may form close to the assay .... Mo

(Prolonged heating with the oxidizing flame is required to obtain the white sublimate, which is yielded satisfactorily only by the sulphide—molybdenite.)

e. White when cold, yellow when hot, light, non-volatile sublimate, which, if moistened with cobalt nitrate and heated, will become bright green when cold................. Zn

(This test should be conducted as follows: Pulverize the material very fine and heat strongly and for some time with the reducing flame; moisten the charcoal where the sublimate has formed, or should form, with a little cobalt nitrate; reheat the assay strongly with the reducing flame, and, if Zn be present in any form but the silicate, enough heat will reach the spot moistened to turn it bright green when cold.)

f. White, heavy sublimate with a blue outer border close to the assay, a yellowish gray coating far from the assay, and a black band between the two. All are easily volatile, burning off with a light bluish green flame, which is also yielded by the assay Te

g. White when cold, yellowish when hot, light,

non-volatile sublimate, which, if moistened with cobalt nitrate and heated, will become a dull bluish green when cold.......... Sn

(This test should be conducted exactly like the one for Zn, described above.)

h. Yellow, volatile sublimate, inclining toward orange when hot, with a very volatile outer fringe of white. Yellow fumes and odor of garlic.... ....As with S

(This result is obtained when a sulphide of As is heated and volatilized too rapidly to permit of complete oxidation. Some of the material is volatilized and deposited as the yellow sulphide of As.)

i. Yellow or orange, non-volatile sublimate, often with a bluish white outer border, deposited very close to the assay...... Pb

j. Yellow or orange, non-volatile sublimate, often with a bluish white outer fringe, deposited very close to the assay...... Bi

(Bi is much rarer than Pb, and the test described above may in the majority of cases be interpreted as indicating Pb. To distinguish with certainty between these two elements, mix the powdered substance with three or four times its volume of "bismuth flux" (equal proportions of potassium iodide and sulphur), and heat on charcoal as usual. The sublimate produced by Bi will be yellow near the assay, but bordered on the outer edge by a brilliant red, which will be missing in the case of Pb.)

k. Brown, volatile sublimate close to the assay,

bordered by a very volatile, heavy white sublimate. Odor of garlic.............. As

(This result is obtained when the material is heated and volatilized too rapidly to admit of complete oxidation of the As. The brown sublimate is metallic As.)

l. Brown, fairly volatile sublimate close to the assay.............................. Cd

m. Reddish brown outer border on a black or steel-gray volatile sublimate. A curious and indescribable, but characteristic, odor and a blue flame...................... Se

n. Lilac or lilac-red, volatile sublimate..Ag with Pb

(Ores yielding this result are rare.)

o. Blue flame and a suffocating pungent odor.. S

(Se, Te, Cu chloride, and other substances burn with a blue flame, but the blue flame combined with the odor is distinctive of S.)

p. Magnetic residue left on charcoal...Fe, Ni, or Co

(If the assay is infusible—does not melt— a magnetic residue indicates Fe.)

II. Treatment on Charcoal with Flux

This test is used when it is desired to reduce a salt to the metallic condition, so the reduction flame is employed.

The substance to be tested is finely pulverized and mixed thoroughly with about three times as much powdered sodium carbonate and a little powdered borax. The mixture is then pressed into a cake at one end of the charcoal and thoroughly fused, beginning at the edges and working toward the center. A little powdered charcoal thoroughly mixed with the material to be fused will often facilitate the reduction. If the assay will not fuse down to a liquid mass, either the amount of flux (sodium carbonate and borax) or of heat applied is deficient. In the latter case, raise the lamp wick and trim off the coal that forms upon it. It is almost useless to hope for satisfactory results until a continuous blast can be blown. Sometimes a fusion that appears to have come to a standstill may be successfully completed by allowing the assay to cool, removing the little cake of semi-fused material with the point of a knife, turning it upside down on the charcoal, and proceeding as at first. Some charcoal usually adheres to the upper surface of the cake and this has a strong reducing influence. This process is always necessary in the case of Sn.

Metallic Sb and many sulphides and arsenides yield metallic globules by this test, but these are distinguishable by their brittleness from those described below.

In addition to sublimates identical with those that form on charcoal without flux, other results are produced which may be thus interpreted:

a. White when cold, yellow when hot, light, non-volatile sublimate, which, if moistened with cobalt nitrate and heated, will become bright green when cold.................. Zn

(Zn compounds which give this test with difficulty or not at all when treated without flux will give a good color when flux is used. The operation should be conducted as described under I.*e.* Care should be exercised not to confuse the *green* Zn coloration that appears *in front of the assay* with a *blue* color *on the assay* itself. The latter will appear whenever fusible material is moistened with cobalt nitrate and heated, regardless of the presence or absence of Zn.)

b. Lilac or lilac-red, moderately volatile sublimate and a white, malleable, metallic button...........................Ag with Pb

(Ores yielding this result are rare.)

c. Yellow or orange, non-volatile sublimate, often with a bluish white outer border, deposited very close to the assay, and a malleable, metallic button, grayish white on a freshly cut surface but oxidizing on exposure to the air........................ Pb

d. Yellow or orange, non-volatile sublimate, often with a bluish white outer border, deposited very close to the assay, and a rather brittle, metallic button, grayish white on a freshly cut surface but oxidizing on exposure to the air...................... Bi

(The button may flatten somewhat when first hammered but is not malleable like Pb. This will usually suffice to distinguish between the two metals, or the test with "bismuth flux," described under I. *j.*, may be applied.)

e. White, malleable, metallic button.......... Ag

(To distinguish between Ag and Pb buttons, place the button to be tested in a small depression on a clean piece of charcoal and heat strongly in the oxidizing flame. If Pb, the characteristic yellow sublimate will form, and, if Ag, there will be no coating or only a faint brownish one. The two metals may also be distinguished after some practice by the fact that the Ag is decidedly harder; the flattened button may be cut only with difficulty while Pb cuts easily.)

f. White, malleable, metallic buttons of small size, which show little or no tendency to coalesce into one large button. A white when cold, yellowish when hot, light, non-volatile sublimate may also form........ Sn

(To distinguish between Sn and Ag, remember that the latter forms one large button, usually, while the many small buttons of the former can be forced to coalesce only with great difficulty and after prolonged blowing. Another method of distinguishing them is to alloy the button in doubt with a somewhat smaller amount of Pb, by melting the two together, and then to note whether the characteristic lilac Ag-Pb sublimate is produced in the oxidizing flame. A third method involves an endeavor to secure the sublimate and cobalt nitrate color reaction of Sn as described under I. *g.*)

g. Yellow, malleable, metallic button......... Au

h. Red, malleable, metallic mass, which it is

impossible to fuse into a single button by means of the blowpipe.................. Cu

i. Gray, malleable, magnetic particles (not globules).......................Fe, Co, or Ni

(The three may be readily distinguished by the "bead tests" (V.), q.v.)

j. The thoroughly fused mass forming the assay, when placed on a clean, moistened silver surface, produces a dark brown or black stain.....................S, Te, or Se

(This test is very delicate, but it must be made carefully in order to succeed. *Exactly* three parts by volume of sodium carbonate must be used for one part of the substance to be tested, and it is best to press the fused mass with a knife or hammer-head against the silver surface which has previously been moistened with a drop of water, and to hold it there for a minute or two. Sometimes a slight stain which may be rubbed off with the fingers or washed off with water is produced. This should be ignored, as the S, Te, or Se stain is permanent.

Unless Te or Se have been detected by tests I.*f*., I.*m*., III.*c*., III.*f*., III.*l*., III.*o*., IV.*g*., or IV.*k*., the presence of a dark stain on the silver may be assumed to indicate S.)

III. Tests in Closed Tubes

It is sometimes desirable to treat the substance just as it is, while for other tests it is best to mix it with three or four times as much sodium carbonate or acid potassium sulphate. In any case, the whole charge should be powdered as fine as possible and enough introduced within a tube to fill it to a height of about half an inch. This may be done with a small paper or tin funnel or chute, or even with a very small knife-blade.

The lower portion of the tube and the charge contained therein should then be heated to redness for some time and the results noted. An alcohol flame alone may be used for this purpose, but the work may be hastened and the results often improved by using the blowpipe on the alcohol or oil flame. Care must be exercised not to use too high a heat or the glass will melt, swell, and break open, or will completely seal up the charge.

Possible results obtainable *without flux* are as follows:

a. Moisture in drops on the walls of the tube a short distance above the charge.......... H_2O

b. Liquid, mirror-like sublimate that collects in globules.............................. Hg

(Only the native metal will give this result without using sodium carbonate flux.)

c. Mirror-like sublimate of large and small, solid, white globules.................... Te

d. Mirror-like sublimate of very small, solid, white globules........................ Cd

(Cd and Te are easily distinguished by using tests I.*f.* and I.*l.*)

e. Mirror-like, solid, black sublimate, often with dull black sublimate above, both volatile As

(A sulphide will not yield this result without using sodium carbonate flux.)

f. White sublimate composed of tiny globules in a narrow zone immediately above the charge, with a ring of yellow globules or liquid at the base, both very slowly volatile.......................... Te

(Tests III.*c.* and III.*f.* are given by different ores of Te or are the result of the application of different temperatures.)

g. White, faint, very slowly volatile sublimate with a little yellow liquid close to the charge................................ Sb

(A sulphide will not yield this result without using sodium carbonate flux.)

h. Reddish liquid when hot, yellow solid when cold; may be almost white when cold if the amount is small...................... S

i. Dark red liquid when hot, orange solid when cold. ....As with S

j. Black when hot, reddish brown when cold, difficultly volatile sublimate ....Sb with S

k. Black, volatile sublimate. If the tube is broken and the sublimate rubbed with a cloth, it will sometimes turn red, but this is unusual.........................Hg with S

l. Black, difficultly volatile sublimate, composed of irregularly shaped drops, liquid when hot. Shades above into a volatile sublimate, reddish brown when hot and dark red when cold . . . . Se

Possible results obtainable with sodium carbonate flux are as follows:

m. Moisture in drops on the walls of the tube a short distance above the charge. . . . . H_2O

n. Liquid, mirror-like sublimate that collects in globules . . . . Hg

o. Mirror-like sublimate of large and small, solid, white globules. . . . . Te

p. Mirror-like sublimate composed of very small, solid, white globules . . . . Cd

(Cd and Te are easily distinguished by using tests I.*f.* and I.*l.*)

q. Mirror-like, solid, black sublimate, often with a dull, solid, black, sublimate above, both volatile . . . . As

r. White, faint, very slowly volatile sublimate with a little yellow liquid close to the assay . . . . Sb

Possible result obtainable with acid potassium sulphate is as follows:

s. After boiling the contents of the tube vigorously for several minutes, the glass immediately above the charge is etched or roughened . . . . F

(This result is most easily recognized by breaking

the tube, washing the interior thoroughly, and then scratching the portion that may be etched with the point of a knife. If etching has occurred, the surface will feel much rougher than does unetched glass.)

IV. Tests in Open Tubes

No flux is used in these tests. The substance to be treated is merely finely powdered, a little of it is placed about half an inch from one end within a tube, and it is then heated strongly while the tube is held as highly inclined as is possible without losing the charge. The alcohol flame may be used, but as good or better results are produced by the oil flame and the blowpipe. In fact, the greater heat of the blowpipe-concentrated oil flame is essential for some of the tests, and this should always be used after securing negative results with the alcohol flame.

The tests are very delicate, but are ordinarily used only to corroborate unsatisfactory charcoal tests.

Possible results may be thus interpreted:

a. Moisture in drops on the walls of the tube.. H_2O

b. Characteristic, suffocating, pungent odor... S
(Many sulphides fail to yield this test.)

c. White, light, very volatile, crystalline sublimate, and odor of garlic................. As

d. White, heavy sublimate, less volatile than

that given by As, and dense white fumes. Slender white crystals may form on the charge if the heat is not too great........ Sb

e. White when cold and light yellow when hot, very heavy, almost non-volatile sublimate and very dense white fumes which pass along the under side of the tube. If a large amount of material be heated very intensely, a little difficultly volatile sublimate, black when hot and reddish brown when cold, may form ....Sb with S

f. White, faint, non-volatile sublimate close to charge ... .Pb with S

g. White, slowly volatile sublimate, which fuses to globules which are yellow hot and colorless cold. Often, yellow globules form around, and a gray sublimate collects above the charge .... Te

h. White, non-volatile sublimate, fusible to yellow drops, lighter when cool ..... .Bi with S

(Bi with S and Te are readily distinguished by tests I.*f.* and I.*j.*, the latter being characteristic of Bi no matter in what combinations it may be.)

i. White, delicate crystals, yellow when hot, form near and over the charge after prolonged heating.......................... Mo

(This result is yielded only by the sulphide.)

j. Yellow, volatile sublimate, inclining toward orange or red when hot, with a very volatile outer border of white. Yellow fumes and odor of garlic . . . . As with S

(This result is often produced when a substance containing S and As is heated too rapidly to allow of complete oxidation.)

k. Black, volatile sublimate where very heavy, which shades toward the upper end of the tube into a volatile coating that is reddish brown when hot and dark red when cold. A curious and indescribable but characteristic odor and lilac fumes, best seen against a black background . . . . Se

l. Black, very volatile sublimate with brownish or iridescent bands and sometimes a mirror-like deposit close to the charge. White fumes, a white volatile sublimate beyond the black, and characteristic odor of garlic. As

(This result is produced when a large amount of material is heated too rapidly to allow of complete oxidation.)

V. Tests with Borax Beads

The beads are made by heating the end of a platinum wire to redness, dipping it into some powdered or granulated borax, reheating the wire and adhering borax, and continuing the process until the bead when thoroughly fused is as large as will remain on the wire. If too small, the color is hard to see, and a bead that is too large will keep falling from the wire. Some of the oxidizing flame tests may be made with the alcohol flame, but all may be secured more quickly and easily with the oil flame and the blowpipe.

The bead should always be kept at the end of the wire, and this may be done by bending the end of the wire and holding it in such a manner that the blowpipe flame is always directed against the side of the bead furthest from the end, forcing it to the end. A loop to hold the bead should not be formed in the wire; it is unnecessary if the wire is clean, and uses up the platinum very rapidly.

Before making any bead test, the material to be tested should be powdered and thoroughly roasted on charcoal. This is accomplished by spreading the powder on the surface of the charcoal in a flat cake, so as to allow free access of air, and heating to a dull red in that part of a small oxidizing flame that is well outside of the blue cone. The reducing flame should then be applied, and, finally, the oxidizing flame used until no odors of As or S are apparent and the assay ceases to burn with a colored flame or to volatilize. Fusion should be prevented if possible. This may be accomplished by mixing the fusible substance with about an equal volume of powdered charcoal, which keeps the particles separated and soon burns away. No satisfactory bead tests need be expected from a substance which volatilizes completely without solidification.

To make the tests, heat the bead as hot as possible and touch it to a very little (a few grains) of the powdered substance to be tested. Some of the latter will adhere to the bead, which should then be heated in the oxidizing flame and any resulting change of color noted. This process should then be repeated on the same bead, using the reducing flame. If the results are negative or too faint to be decisive, more of the powder should be dissolved in the bead and the process continued until a relatively large amount has been added, when the bead is said to be saturated.

The colors of the beads are due to the presence of oxides of the various elements, and these oxides vary greatly in their coloring powers. In some cases, deep, vivid colors are obtained from a few grains of the powdered substance, while in other instances it is necessary to dip the bead many times into the powdered substance before the characteristic colors are produced. Beads in the former class may be considered saturated when the color is so deep as to make them practically opaque; those in the latter class are saturated when they refuse to absorb more of the material. This may require half a dozen applications of the powder.

In case a bead becomes so saturated upon the first application to the powder that its color is indeterminate, it may be flattened upon the anvil while still warm and the color readily observed in the thin cake thus formed. This cake may then be broken into many pieces, and a few of these added to a fresh bead without saturating the latter. All beads, even when saturated, should be perfectly clear unless otherwise noted. If this is not the case, a higher heat should be applied or new beads formed, as a bead that has been worked with for some time in different flames is apt to become translucent or opaque.

It should be remembered that a bead containing incompletely roasted powder is very apt to be brown in both flames, and then prolonged heating is required to expel the S or other interfering element or elements.

The bead test should never be used on a substance suspected to contain Cu, as that element will alloy with the platinum and give Cu beads whenever the same wire is used in subsequent tests. Other elements are apt to alloy with the platinum, particularly during the formation of reduction beads, and these will make the wire brittle and cause it to break easily, but will not interfere with tests made before the break occurs.

To remove a bead from the wire, either break it off on the anvil or jar it off while in a molten condition. When the latter method is used, it is a good idea to save the beads obtained from known substances and compare them with those given by unknown materials.

The most difficult operation involved in making the bead tests is the production of a good, continuous reducing flame, yet this is very important and must be mastered. Manganese gives a very highly colored bead in the oxidizing flame, and the production of the colorless bead in the reducing flame is a good test of ability in this line.

If a substance contains two elements each of which yields a characteristic bead, one may so modify the other as to give intermediate results. However, in most cases of this kind, one color will completely mask the other, and this makes it then impossible to detect both elements.

The accompanying table indicates the colors of the borax and salt of phosphorus beads (see below) yielded by the elements named in both oxidizing and reducing flames. The following abbreviations are used: O.F. = oxidizing flame. R.F. = reducing flame. W = warm bead. W* = saturated warm bead. C = cold bead. C* = saturated cold bead.

As an illustration, consider iron. The table shows that iron gives in the oxidizing flame a borax bead that is yellow while warm and colorless when cold unless the bead is saturated; then, it is yellow when cold. In the reducing flame, the bead is green when warm and colorless when cold unless saturated, when it is bottle-green when cold. Likewise, both the non-saturated and saturated salt of phosphorus beads are yellow when warm and

TABLE OF BORAX BEAD TESTS

		Colorless.	Yellow.	Green.	Blue.	Violet.	Red.	Brown.	References.
a. Iron	O.F.	C	W*WC*						NOTE 1
	R.F.	C		W*WC*					
b. Molybdenum	O.F.	C*C	W*W						NOTE 2
	R.F.	WC						W*C*	
c. Titianium	O.F.	C	W*WC*						
	R.F.	C	W*WC*						
d. Tungsten	O.F.	WC*C	W*						
	R.F.	WC*C	W*						
e. Uranium	O.F.	C	W*WC*						
	R.F.	C	W*W	C*					
f. Vanadium	O.F.	C	W*WC*						NOTE 3
	R.F.			W*WC*C					
g. Chromium	O.F.		W	C*C			W*		NOTE 4
	R.F.			W*WC*C					
h. Copper	O.F.			W*W	C*C				NOTE 5
	R.F.			W*W	C*C		C*		
i. Cobalt	O.F.				W*WC*C				NOTE 6
	R.F.				W*WC*C				
j. Manganese	O.F.					W*WC*C			NOTE 7
	R.F.	W*WC*C							
k. Nickel	O.F.					W*W		C*C	NOTE 8
	R.F.	WC							

Notes on Table of Borax Bead Tests

Note 1. The strongly saturated bead is a dull bottle- or olive-green when cold after treatment with the reducing flame. A very small quantity of ore should be used for the unsaturated bead tests.

Note 2. The strongly saturated bead is opaque after treatment with the reducing flame.

Note 3. Saturated and non-saturated warm beads are greenish yellow after treatment in the oxidizing flame.

Note 4. The cold, non-saturated bead is yellowish green after treatment with the oxidizing flame. A small quantity of ore should be used for the unsaturated bead tests.

Note 5. The strongly saturated bead may be opaque and dark brownish red after treatment in the reducing flame. A very small quantity of ore should be used for the unsaturated bead tests.

Note 6. A very small quantity of ore should be used for the unsaturated bead tests.

Note 7. An extremely small quantity of ore should be used for the unsaturated bead tests.

Note 8. The saturated bead is gray and opaque after treatment with the reducing flame.

Notes on Table of Salt of Phosphorus Bead Tests

Note 1. The warm bead is pale to deep yellowish green (depending upon the degree of saturation) after treatment with the reducing flame. A very small quantity of ore should be used for the unsaturated bead tests.

Note 2. The cold saturated bead is very pale violet after treatment with the reducing flame.

Note 3. The cold saturated bead is greenish blue after treatment with the reducing flame.

Note 4. The green bead obtainable with the reducing flame cannot be reoxidized to yellow.

Note 5. A small quantity of ore should be used for the unsaturated bead tests.

Note 6. The cold non-saturated bead is pale green after treatment with the reducing flame.

Note 7. The strongly saturated bead may be opaque and dark brownish red after treatment in the reducing flame. A very small quantity of ore should be used for the unsaturated bead tests.

Note 8. A very small quantity of ore should be used for the unsaturated bead tests.

Note 9. An extremely small quantity of ore should be used for the unsaturated bead tests.

TABLE OF SALT OF PHOSPHORUS BEAD TESTS.

		Colorless.	Yellow.	Green.	Blue.	Violet.	Red.	Brown.	References.
a. Iron	O.F. R.F.	C*C C	W*W	W*W				C*	NOTE 1
b. Titanium	O.F. R.F.	C*C C	W*W W*W			C*			NOTE 2
c. Tungsten	O.F. R.F.	WC*C WC	W* W*		C*				NOTE 3
d. Vanadium	O.F. R.F.		W*WC*C	W*WC*C					NOTE 4
e. Nickel	O.F. R.F.		C*C C*C				W*W W*W		
f. Chromium	O.F. R.F.		W	W*C*C W*WC*C					NOTE 5
g. Uranium	O.F. R.F.		W*W	C*C W*WC*C					
h. Molybdenum	O.F. R.F.	C		W*WC* W*WC*C					NOTE 6
i. Copper	O.F. R.F.			W*W W*W	C*C C*C		C*		NOTE 7
j. Cobalt	O.F. R.F.				W*WC*C W*WC*C				NOTE 8
k. Manganese	O.F. R.F.	W*WC*C				W*WC*C			NOTE 9

colorless when cold in the oxidizing flame, while in the reducing flame the non-saturated salt of phosphorous bead is pale yellowish green when warm and colorless when cold, and the saturated bead is deep yellowish green when warm and brown when cold.

VI. Tests with Salt of Phosphorus (Sodium Ammonium Phosphate or Microcosmic Salt) Beads

These tests are made in exactly the same manner as are those with borax beads and the same precautions should be observed. The salt of phosphorus is, however, much more liquid than borax, especially when first heated, and drops off the wire very easily. It will be found necessary to build the bead up gradually by the addition of small particles picked up on the hot wire one after the other and to use smaller beads than with borax. If difficulty in retaining the bead is still experienced, it will be found advisable to form a loop in the end of the wire by bending it around the point of a lead pencil. In any case, it is best while forming the bead to allow the flame to play upon the under side of the fusing mass, thus buoying it upward and decreasing its tendency to drop off.

The salt of phosphorus tests are necessary for

the recognition of some of the elements, but in most cases they will be found useful merely to corroborate unsatisfactory borax bead tests, and they may often be omitted.

The accompanying table indicates the colors of both salt of phosphorus and borax beads in both oxidizing and reducing flames. The abbreviations and use of the table have been explained in the discussion of the tests with borax beads.

VII. Flame Tests

When volatilized, certain substances impart more or less decided colors to a flame, and the recognition of these colors constitutes distinctive tests for such elements.

The flames are best seen in a dark room or against a dark background, and even then it requires close application to perceive the very brief flashes of color which constitute the tests in many instances.

Four different methods of making these tests may be used, and they should be applied in the order given, it being unnecessary, however, to seek further results after a determination has been made by any of the methods. If two elements, both of which yield characteristic flames, are present, one will usually so mask the other as to make the recognition of both impossible.

The second, third, or fourth method should never be used upon a substance with a metallic lustre without a thorough preliminary roasting, as As, Sb, Pb, and other easily reduced elements, are apt to form fusible alloys with the platinum wire or forceps and thus ruin them.

Some elements yield their characteristic flame colors best at a low heat while others require the highest heat available, so it is always best to use both low and high temperatures for each test.

Some substances which ordinarily yield no flame tests may have some of their constituents converted into volatile, flame-tinting compounds by treatment with some reagent, usually HCl or H_2SO_4. In fact, it is a safe practice always to dampen the material to be tested with HCl, not even trying a test without the use of that acid. If no tinted flame results, H_2SO_4 should be tried.

First Method. Place a fragment or some of the powdered substance upon charcoal, moisten with a few drops of concentrated HCl, and heat in the hottest portion of the blowpipe flame. The only result that need be sought and noted (the others being more easily obtained, or the elements being more easily recognized, by other methods) is the following:

a, Azure-blue flame, with or without flashes, or a border, of emerald green............ **Cu**

(If this test results favorably, care should be taken **not** to make the bead tests, which could only yield Cu, or to heat the material in the platinum forceps, since in either of these operations the Cu will alloy with the Pt and ruin it.)

Second Method. (This is applicable only to carbonates. See IX.*b.*) Seize a sliver of considerable size in the platinum forceps, moisten it with *dilute* HCl (one part of acid to three or four parts of water), and hold the particle near the base of the flame of an alcohol lamp. the blowpipe flame not being used for this test. If the result is negative, repeat the operation, but use the hot tip of the alcohol flame instead of the base. In a few cases, cold or hot concentrated acid is required and these should be tried as a last resort, but the dilute acid should be first tried, as there are several substances which will not give a good reaction with the strong acid.

The splinter should be moistened by immersing it in the acid and holding it there until there is a vigorous effervescence (evolution of gas). Pure, fresh alcohol and acids should be used for these tests, as otherwise a yellow Na flame is apt to be very prominent and may mask the flames yielded by Ba or Pb.

If an alcohol lamp is not available, all of the tests described below may be obtained by using the method next given, but the results are more vivid and are easier obtained in the manner just described, if the substance tested is a carbonate.

Possible results may be thus interpreted:

b. Scarlet flame, lilac through blue glass....... Sr

(A Li flame might easily be confused with that from Sr, but no known Li carbonate occurs in nature, so this test when obtained in the above described manner always indicates the presence of Sr.)

c. Yellowish red flame, greenish through blue glass.................................. Ca

(This is difficult to distinguish at first from the Sr flame, but is considerably less vivid and is inclined toward orange. It may be positively identified by the fact that Ca salts give good tests no matter whether dipped in concentrated or dilute acid and *dilute* acid *must* be used to secure a good Sr flame.)

d. Yellow flame.............................. Na

(This test is too delicate to be used with safety unless very intense and persistent. Specimens that have been handled will become sufficiently charged with Na from the fingers to give a good flame test.)

e. Yellowish green flame, pale tint............ Ba

(Masked by strong Na flame.)

f. Blue flame, pale tint....................... Pb

(Hot, concentrated acid is required for this test, and even then the color will appear for only an instant when the splinter is held in the tip of the flame.)

Note. A vivid Cu flame may be obtained in this manner, but this should never be attempted, as the forceps will thereby be ruined.

Third Method. Hold an extremely fine splinter

(as slender as a very fine needle) in the platinum forceps, moisten it with a drop of HCl, and introduce it into the hottest part of the blowpipe flame. A flash or a continuous appearance of color may be imparted to the flame. If the result is negative or unsatisfactory, remoisten the splinter and reheat. If this fails, repeat the operation with H_2SO_4 instead of HCl. Should the splinter decrepitate, try heating it very slowly, and, if it still flies to pieces, the fourth method must be used.

Possible results may be thus interpreted:

g. Carmine flame, violet through blue glass..... Li

h. Scarlet flame, lilac through blue glass........ Sr

(The colors of the Li and Sr flames are so similar that they are easily confused, but they may be readily distinguished by the fact that a substance that has been ignited and has given a Sr flame will turn moist red litmus paper blue when crushed and placed upon it. Li minerals show no effect of this kind.)

i. Yellowish red flame, greenish through blue glass.................................... Ca

(This is difficult to distinguish at first from the Sr or Li flames, but is considerably less vivid than either and is inclined toward orange.)

j. Yellow flame .... Na

(This test is too delicate to be used with safety unless very intense and persistent. Specimens that have been handled will become sufficiently charged with Na from the fingers to give a good flame test.)

k. Yellowish green flame, pale tint.............. Ba

(Masked by strong Na flame.)

l. Bluish green flame, pale tint................. P

(H_2SO_4 must be used for this test and the result is not usually very satisfactory.)

m. Bright green flame......................... B

(B minerals which do not give a flame test in this way should be powdered and mixed thoroughly with about three volumes of a mixture of equal parts of powdered acid potassium sulphate and calcium fluoride—fluorite. This should be introduced into the flame as described in the fourth method.)

n. Blue flame, pale tint....................... Pb

o. Blue flame, pale tint........................ Sb

(The reducing flame must be used to obtain this result, which is neither very satisfactory nor determinative. Care should be taken not to test an Sb ore in this way, as it is apt to alloy with, and ruin, the platinum forceps.)

p. Violet flame, violet through blue glass...... K

(This is difficult to obtain in most cases and is entirely masked by a pale Na flame. The latter is, however, entirely absorbed by blue glass, which transmits the K flame as of lilac or violet color, depending upon the shade of the glass.)

Note. A vivid Cu flame may be obtained in this manner, but this should never be attempted, as the forceps will thereby be ruined.

Fourth Method. Powder the material to be tested very fine, pick up a little of this powder upon a flattened platinum wire moistened with HCl, and introduce the powder-coated wire into the hottest part of the blowpipe flame. Momentary flashes or continuous appearances of color may be imparted to the flame, which are to be interpreted as under the third method. If the results are negative, repeat the operation, using H_2SO_4 instead of HCl.

This method is not usually very satisfactory and should not be used unless it is impossible to obtain a splinter fine enough to be tested by the third method. Minerals which decrepitate badly, which are very soft, or which occur as a powder must, of course, be tested by this last method.

VIII. Cobalt Nitrate Coloration Tests

To make these tests, hold a small splinter of the substance to be tested in the platinum forceps and heat it in the blowpipe flame to the highest possible temperature. Then examine it with a lens; if it shows any signs of fusion, this test cannot be applied. If non-fusible, moisten it with cobalt nitrate and ignite strongly in the hottest part of the blowpipe flame. It will first turn black but after prolonged heating may assume a characteristic tint. If a splinter of the substance cannot be obtained, it should be powdered and the test conducted upon a flat cake of the powder upon charcoal. Longer heating is required by this method, however, and the results are not apt to be as satisfactory.

This test can be applied only to non-fusible, white or faintly tinted minerals, or those which become white or faintly tinted upon ignition.

Possible results may be thus interpreted:

a. Blue coloration.................Al or Zn silicate

(Al minerals and Zn silicate give identical results by this test. If Zn has been obtained by tests I.*e.*, or II.*a.*, it is impossible to test for Al in this way. If Zn is not present, this test may be interpreted as indicating the presence of Al. Occasionally a little green forms with the blue on Zn silicates, which never happens in the case of Al minerals.)

b. Green coloration, dark.................... Sb

c. Green coloration, bright tint, best seen when cold .... Zn

d. Pinkish or flesh-tint coloration............ Mg

IX. Tests with Acids

These are really purely chemical and should not be included under blowpipe analysis, but two are so simple and useful that they are given below.

a. The finely powdered material, when boiled almost to dryness in concentrated nitric acid, yields a gelatinous mass........a silicate

b. The powdered material effervesces vigorously

when placed in a test-tube containing some condition of hydrochloric acid......a carbonate

(In some cases the tests appear in cold, dilute acid, while, in other instances, hot dilute, cold concentrated, or hot concentrated may be required. It is best to experiment with the dilute acid first, and then, if results are negative, to increase the temperature until all conditions of acid have been tried.

It is not always necessary to powder the material to be tested, but this is sometimes required and never does any harm.

Care should be taken not to confuse effervescence (escape of CO_2) with boiling (escape of steam) when boiling acid is used.

Some sulphides may effervesce in hydrochloric acid, but these may be distinguished from carbonates by the fact that they yield a gas (H_2S) that smells like bad eggs.)

CHAPTER III

OUTLINE FOR QUALITATIVE BLOWPIPE ANALYSIS

Blowpipe analysis is ordinarily used for the purpose of ascertaining what elements an unknown substance contains, this process being known in chemistry as qualitative analysis. It is not possible, excepting in a few cases and by the application of very refined methods not here discussed, to determine how much of an element is present—to make a quantitative analysis of the substance; but, if the elements present are known, it is usually possible to determine the nature of the substance by applying the principles set forth in Chapter V. Where the nature of a mineral has thus been determined, it is comparatively easy to compute the percentage composition with considerable accuracy in many cases, by applying the principles presented in Chapter VI, A tabulation of the principal elements present in many common minerals are given in Chapter V. Even when the name and nature of the mineral cannot be ascertained by blowpipe methods, the ability to ascertain the commercially important elements that it contains is often of the greatest value, since it is then possible to decide whether it is worth while to procure a chemical analysis or an assay of the substance.

The following scheme has been devised as a guide for making a *complete* qualitative analysis of an unknown substance with the greatest economy of time and labor. It should not be expected that every mineral containing As, for instance, will yield every test for As mentioned in Chapter II or in the following outline, but in most cases the mineral will give one or more of the tests there given. It is poor practice to assume that a mineral contains a certain element and then test for that element, repeating the operation for other elements. A far better plan is to follow the outline rigidly and to draw conclusions from the results secured. This saves time and makes it impossible to forget to try tests.

If it is suspected that Hg is present in a substance, it should be tested only in the closed tube (see III.*b*.), as the vapors are very poisonous.

Outline

A. Powder material very fine and place upon charcoal, pressing out with a knife-blade into a flat cake. Moisten with water if necessary to keep the charge from flying off the charcoal.

1. Test for As, Sb, Sb with As, Mo, Te, As with

S, Pb, Bi, Cd, Se, Ag with Pb, and S. (See I.)

2. Test for Fe. (See I.*p*.)

Separate the residue from the above tests into two portions and save one of these for a later test. On the portion still remaining on the charcoal make:

3. Test for Zn and Sn. (See I.*e*. and I.*g*.)
4. Test for Cu on residue from A.3. (See VII.*a*.)
5. Test for Al, Mg, Zn, Sb, and Zn silicate on residue from A.4. (See VIII.)

(This test can be made only upon light colored, infusible material, and need not be tried if Zn has already been found.)

B. If A.4. gave no Cu flame (if Cu is present it is useless to make the following tests) make borax and salt of phosphorus bead tests for the elements named below, using the residue saved from A.2.

1. Test for Fe, Mo, Ti, W, U, V, Cr, Cu, Co, Mn, and Ni. (See V. and VI.)

C. In case A.1. gave a result that might be interpreted as indicating either Pb or Bi, use the bismuth flux test to distinguish them.

1. Test for Bi on some of the original material, using bismuth flux. (See I.*j*.)

D. If A.4. gave no Cu flame (if Cu is present it is useless to make the following tests), hold a piece of the original material as large as a toothpick or match in the platinum forceps and make the flame tests as described under VII., Second Method.

1. Test for Sr, Ca, Na, Ba, and Pb carbonates. (See VII., Second Method.)

   If no satisfactory results are obtained from D.1., make the flame tests as described under VII., Third Method.

2. Test for Li, Sr, Ca, Na, Ba, P, B, Pb, Sb, and K. (See VII., Third Method.)

   If no satisfactory results are obtained from D.2., make the flame tests as described under VII., Fourth Method.

3. Test for Li, Sr, Ca, Na, Ba, P, B, Pb, Sb, and K. (See VII., Fourth Method.)

E. On some of the original material, powdered, make the closed tube tests without flux.

1. Test for H_2O, Hg, Te, Cd, As, Sb, S, As with S, Sb with S, Hg with S, and Se. (See III.*a.* to *l.*)

   On some of the original material, powdered, make the closed tube tests with flux, provided E.1. has not yielded determinative results.

2. Test for H_2O, Hg, Te, Cd, As, and Sb. (See III.*m.* to *r.*)

   On some of the original material, powdered, make the closed tube test with acid potassium sulphate.

3. Test for F. (See III.*s*.)

F. On some of the original material, powdered, make the open tube tests.

1. Test for H_2O, S, As, Sb, Sb with S, Pb with S, Te, Bi with S, Mo, As with S, and Se. (See IV.)

Note. It is always well to make tests E. and F. even when A. has given determinative results, as several of the elements determined by means of tests E. and F. may fail to show in test A. It is a good practice to corroborate results obtained by test A. by means of tests E and F.

G. On some of the original material, powdered, make the tests on charcoal with flux.

1. Test for Zn, Ag with Pb, Pb, Bi, Ag, Sn, Au, Cu, and Fe, Co, or Ni. (See II.)

   On the residue from G.1., make the silver test for S, Te, or Se, provided these elements, or one of them, have not already been detected by other tests.

2. Test for S, Te, or Se. (See II.*j*.)

H. On some of the original material, powdered, make the tests with acids in test-tubes.

1. Test for a silicate with nitric acid. (See IX.*a*.)
2. Test for a carbonate with hydrochloric acid. (See IX.*b*.)

CHAPTER IV

INDEX TO ALL OF THE TESTS YIELDED BY THE VARIOUS ELEMENTS

It is the purpose of this index to furnish a complete list of all the blowpipe tests for any element discussed in the preceding pages. It will be found useful where the interest is concentrated upon one or two elements to the exclusion of all others, but should not be used in making a complete qualitative blowpipe analysis of a substance. It should not be expected, that any mineral will necessarily yield *all* of the tests for each of the constituents.

The references here given are to the page or pages on which tests for the various elements are described.

Cadmium: 15 and 20.
Calcium: 36 and 37.
Chromium: 30 and 31.
Cobalt: 15, 19, 30, and 31.
Copper: 19, 30, 31, and 34.
Flüorine: 22.
Gold: 18.
Iron: 15, 19, 30, and 31.
Lead: 14, 15, 17, 24, 36, and 38.
Lithium: 37.
Magnesium: 40.
Manganese: 30 and 31.
Mercury: 20, 21, and 22.
Molybdenum: 13, 24, 30, and 31.
Nickel: 15, 19, 30, and 31.
Phosphorus: 38.
Potassium: 38.
Selenium: 15, 19, 22, and 25.
Silver: 15, 17, and 18.
Sodium: 36 and 37.
Strontium: 36 and 37.
Sulphur: 14, 15, 19, 21, 23, 24, and 25.
Tellurium: 13, 19, 20, 21, 22, and 24.
Tin: 14 and 18.
Titanium: 30 and 31.
Tungsten: 30 and 31.
Uranium: 30 and 31.

Zinc: 13, 17, and 40.
Water: 20, 22, and 23.
A carbonate: 41.
A silicate: 40.

CHAPTER V

THE DETERMINATION OF MINERALS BY MEANS OF THE BLOWPIPE

WHILE the determination of the constituent elements of a mineral is the usual aim of a blowpipe analysis, it is often desirable to be able to assign the correct mineralogical name to the substance, since, when this can be done, it is frequently possible to determine its percentage composition with considerable accuracy. It is not, unfortunately, possible in many cases to determine a mineral by blowpipe tests alone; these must be considered in connection with the physical characteristics before a reliable decision as to the correct name can be safely made. There are, however, a considerable number of minerals with unique groups of constituents which may be determined by blowpipe analyses, and it is hoped that the following table will prove useful in this respect. It includes most of the important ores and some of lesser importance, as well as a number of gangue minerals, but some very important ores are omitted as well as a great number of common minerals for the reason that blowpipe tests are little or no aid in their recognition. In a large number of cases two or more minerals on this table yield tests for identically the same elements. Such substances can be readily distinguished by their appearance or by simple physical tests. For this purpose almost any book on mineralogy will answer, but the author naturally prefers his own work, A Pocket Handbook of Minerals, published by John Wiley & Sons, New York, which places all the emphasis upon the physical distinctions.

In the following table the chemical formula is placed in parentheses after the name of each mineral. By applying the principles presented in Chapter VI it should be a comparatively simple matter to compute the percentage of any or all elements present in any mineral of fixed composition. The letters to the left of each name are the symbols of the elements that may be found by means of the blowpipe; in comparatively few cases do they constitute all the elements present in the mineral before which they stand, but the remainder fail to give satisfactory tests with the blowpipe.

The name of a mineral is repeated under each element for which it yields blowpipe tests. Water (H_2O) is treated as an element.

Aluminum.

Al—Corundum (Al_2O_3)

Al, Li—Spodumene ($LiAlSi_2O_6$).

Al, P, H_2O—Turquois ($AlPO_4.Al(OH)_3+H_2O$ $+Cu$).

Al, K—Orthoclase ($KAlSi_3O_8$).
Al,S,H_2O—Alunite ($K_2SO_4.3Al_2O_3.3SO_3.6H_2O$).
Al, H_2O—Bauxite ($Al_2O_3+2H_2O$).
Al, H_2O—Kaolin ($Al_2Si_2O_7+2H_2O$).

Antimony.

Sb—Native Antimony (Sb).
Sb—Cervantite (Sb_2O_4).
Sb, As—Allemontite (SbAs).
Sb, Cu, S—Tetrahedrite ($Cu_8Sb_2S_7$).
Sb, Pb, S—Jamesonite ($Pb_2Sb_2S_5$).
Sb, Ag, S—Pyrargyrite (Ag_3SbS_3).
Sb, Ag, S—Stephanite (Ag_5SbS_4).
Sb, S—Stibnite (Sb_2S_3).

Arsenic.

As—Native Arsenic (As).
As, Sb—Allemontite (AsSb).
As, Co—Smaltite ($(Co,Ni)\ As_2$).
As, Co, S—Cobaltite (CoAsS).
As, Cu, S—Tennantite ($Cu_8As_2S_7$).
As, Cu, S—Enargite (Cu_3AsS_4).
As, Cu, H_2O—Olivenite ($Cu_4As_2O_9+H_2O$).
As, Cu, H_2O—Conichalcite ($(Cu,Ca)_4As_2O_9+3/2H_2O$).
As, Fe—Lollingite ($FeAs_2$).
As, Fe, S—Arsenopyrite (FeAsS).
As, Ni—Niccolite (NiAs).
As, Pb—Mimetite ($PbCl_2.3Pb_3As_2O_8.$).
As, Ag, S—Proustite (Ag_3AsS_3).

As, S—Realgar (As_2S_2).
As, S—Orpiment (As_2S_3).

Barium.

Ba—Witherite ($BaCO_3$).
Ba, S—Barite ($BaSO_4$).

Bismuth.

Bi—Native Bismuth (Bi).
Bi—Bismite (Bi_2O_3).
Bi, S—Bismuthinite (Bi_2S_3).
Bi, Te—Tetradymite (TeBi).
Bi, H_2O—Bismutite ($Bi_2CO_5 + H_2O$).

Boron.

B—Boracite ($Mg_7Cl_2B_{16}O_{30}$).
B, Na, H_2O—Borax ($Na_2B_4O_7 + 10H_2O$).
B, H_2O—Colemanite ($Ca_2B_6O_{11} + 5H_2O$).

Cadmium.

Cd, S—Greenockite (CdS).

Calcium.

Ca—Calcite ($CaCO_3$).
Ca, F—Fluorite (CaF_2).
Ca, Mg—Dolomite ($(Ca, Mg)\ CO_3$).
Ca, P—Apatite ($Ca\ (Cl, F)_2 . 3Ca_3P_2O_8$).
Ca, S—Anhydrite ($CaSO_4$).
Ca, S, H_2O—Gypsum ($CaSO_4 + 2H_2O$).

Chromium.

Cr, Fe—Chromite ($FeCr_2O_4$).
Cr, Pb—Crocoite ($PbCrO_4$).

Cobalt.

Co, As—Smaltite ((Co, Ni) As_2).

Co, As, S—Cobaltite (CoAsS).

Copper

Cu—Native Copper (Cu).

Cu—Cuprite (Cu_2O).

Cu—Tenorite (CuO).

Cu—Atacamite ($CuCl_2.3Cu(OH)_2$).

Cu, Sb, S—Tetrahedrite ($Cu_8Sb_2S_7$).

Cu, As, S—Tennantite ($Cu_8As_2S_7$).

Cu, As, S—Enargite (Cu_3AsS_4).

Cu, As, H_2O—Olivenite ($Cu_4As_2O_9+H_2O$).

Cu, As, H_2O—Conichalcite ($(Cu, Ca)_4As_2O_9+3/2H_2O$).

Cu, Fe, S—Bornite (Cu_5FeS_4).

Cu, Fe, S—Chalcopyrite ($CuFeS_2$).

Cu, S—Chalcocite (Cu_2S).

Cu, S—Covellite (CuS).

Cu, S, H_2O—Chalcanthite ($CuSO_4+5H_2O$).

Cu, H_2O—Malachite ($Cu_2CO_4+H_2O$).

Cu, H_2O—Azurite ($Cu_3C_2O_7+H_2O$).

Cu, H_2O—Chrysocolla ($CuSiO_3+2H_2O$).

Fluorine.

F, Ca—Fluorite (CaF_2).

F, Na—Cryolite (Na_3AlF_6).

Gold.

Au—Native Gold (Au).

Au, Te—Calaverite ($AuTe_2$).

Iron.

Fe—Hematite (Fe_2O_3).
Fe—Magnetite (Fe_3O_4).
Fe—Siderite ($FeCO_3$).
Fe, As—Lollingite ($FeAs_2$).
Fe, As, S—Arsenopyrite (FeAsS).
Fe, Cr—Chromite ($FeCr_2O_4$).
Fe, Mn, Zn—Franklinite ($(Fe, Zn, Mn)_3O_4$).
Fe, P, H_2O—Vivianite ($Fe_3P_2O_8+8H_2O$).
Fe, S—Pyrrhotite (Fe_nS_{n+1}).
Fe, S—Pyrite (FeS_2).
Fe, Ti—Ilmenite ($FeTiO_3$).
Fe, H_2O—Limonite ($2Fe_2O_3+3H_2O$).

Lead.

Pb—Cerussite ($PbCO_3$).
Pb, Sb, S—Jamesonite ($Pb_2Sb_2S_5$).
Pb, As—Mimetite ($PbCl_2.3Pb_3As_2O_8.$).
Pb, Cr—Crocoite ($PbCrO_4$).
Pb, Mo—Wulfenite ($PbMoO_4$).
Pb, P—Pyromorphite ($PbCl_2.3Pb_3P_2O_8$).
Pb, S—Galenite (PbS).
Pb, S—Anglesite ($PbSO_4$).
Pb, V—Vanadinite ($PbCl_2.3Pb_3V_2O_8$).
Pb, V, etc.—Uraninite (?).

Lithium.

Li, Al—Spodumene ($LiAlSi_2O_6$).

Magnesium

Mg—Magnesite ($MgCO_3$).

Mg, Ca—Dolomite ((Mg, Ca)CO_3).
Mg; H_2O—Talc ($Mg_3Si_4O_{11}+H_2O$).

Manganese.

Mn—Pyrolusite (MnO_2).
Mn—Rhodochrosite ($MnCO_3$).
Mn—Rhodonite ($MnSiO_3$).
Mn, Fe, Zn—Franklinite (($(Mn, Fe, Zn)_3O_4$).
Mn, S—Alabandite (MnS).
Mn, H_2O—Manganite ($Mn_2O_3+H_2O$).
Mn, H_2O—Psilomelane (MnO_2+2H_2O).

Mercury.

Hg—Native Mercury (Hg).
Hg, S—Cinnabar (HgS).

Molybdenum.

Mo—Molybdite (MoO_3).
Mo, Pb—Wulfenite ($PbMoO_4$).
Mo, S—Molybdenite (MoS_2).

Nickel.

Ni, As—Niccolite (NiAs).
Ni, S—Millerite (NiS).
Ni, H_2O—Garnierite ($H_2(Ni, Mg)SiO_4+H_2O$).

Phosphorus.

P, Al, H_2O—Turquois ($AlPO_4.Al(OH)_3+H_2O+Cu$).
P, Ca—Apatite ($Ca(Cl, F)_2.3Ca_3P_2O_8$).
P, Fe, H_2O—Vivianite ($Fe_3P_2O_8+8H_2O$).
P, Pb—Pyromorphite ($PbCl_2.3Pb_3P_2O_8$).

Potassium.

K—Sylvite (KCl).
K, Al—Orthoclase ($KAlSi_3O_8$).

Selenium.

Se, Pb—Clausthalite (PbSe).

Silver.

Ag—Native Silver (Ag).
Ag—Cerargyrite (AgCl).
Ag, Sb, S—Pyrargyrite (Ag_3SbS_3).
Ag, Sb, S—Stephanite (Ag_5SbS_4).
Ag, As, S—Proustite (Ag_3AsS_3).
Ag, Te—Petzite ($(Ag, Au)_2Te$).
Ag, Te—Sylvanite ($(Ag, Au)Te_2$).
Ag, S—Argentite (Ag_2S).
Ag, Te—Hessite (Ag_2Te).

Sodium

Na—Halite (NaCl).
Na, B, H_2O—Borax ($Na_2B_4O_7+10H_2O$).
Na, F—Cryolite (Na_3AlF_6).
Na, S—Thenardite (Na_2SO_4).
Na, H_2O—Natron ($Na_2CO_3+10H_2O$).
Na, H_2O—Trona ($Na_2CO_3.HNaCO_3+2H_2O$).

Strontium

Sr—Strontianite ($SrCO_3$).
Sr, S—Celestite ($SrSO_4$).

Sulphur.

S—Native Sulphur (S).
S,Al,H_2O—Alunite($K_2SO_4.3Al_2O_3.3SO_3.6H_2O$).
S, Sb—Stibnite (Sb_2S_3).
S, Sb, Cu—Tetrahedrite ($Cu_8Sb_2S_7$).
S, Sb, Pb—Jamesonite ($Pb_2Sb_2S_5$).
S, Sb, Ag—Pyrargyrite (Ag_3SbS_3).
S, As—Realgar (As_2S_2).
S, As—Orpiment (As_2S_3).
S, As, Cu—Tennantite ($Cu_8As_2S_7$).
S, As, Cu—Enargite (Cu_3AsS_4).
S, As, Fe—Arsenopyrite (FeAsS).
S, As, Ag—Proustite (Ag_3AsS_3).
S, Ba—Barite ($BaSO_4$).
S, Bi—Bismuthinite (Bi_2S_3).
S, Cd—Greenockite (CdS).
S, Ca—Anhydrite ($CaSO_4$).
S, Ca, H_2O—Gypsum ($CaSO_4 + 2H_2O$).
S, Cu—Chalcocite (Cu_2S).
S, Cu—Covellite (CuS).
S, Cu, Fe—Bornite (Cu_5FeS_4).
S, Cu, Fe—Chalcopyrite ($CuFeS_2$).
S, Cu, H_2O—Chalcanthite ($CuSO_4 + 5H_2O$).
S, Fe—Pyrrhotite (Fe_nS_{n+1}).
S, Fe—Pyrite (FeS_2).
S, Pb—Galenite (PbS).
S, Pb—Anglesite ($PbSO_4$).
S, Mn—Alabandite (MnS).
S, Mo—Molybdenite (MoS_2).

S, Ni—Millerite (NiS).
S, Ag—Argentite (Ag_2S).
S, Na—Thenardite (Na_2SO_4).
S, Sr—Celestite ($SrSO_4$).
S, Zn—Sphalerite (ZnS).

Tellurium.

Te—Native Tellurium (Te).
Te, Bi—Tetradymite (BiTe).
Te, Au—Calaverite ($AuTe_2$).
Te, Ag—Petzite ($(Au, Ag)_2Te$).
Te, Ag—Hessite (Ag_2Te).
Te, Ag—Sylvanite ($(Au, Ag)Te_2$).

Tin.

Sn—Cassiterite (SnO_2).

Titanium.

Ti—Rutile (TiO_2).
Ti—Titanite ($CaTiSiO_5$).
Ti, Fe—Ilmenite ($FeTiO_3$).

Tungsten.

W—Scheelite ($CaWO_4$).

Uranium.

U, Pb, etc.—Uraninite (?).

Vanadium.

V, Pb—Vanadinite ($PbCl_2.3Pb_3V_2O_8$).

Zinc.

Zn—Zincite (ZnO).
Zn—Smithsonite ($ZnCO_3$).

Zn—Willemite (Zn_2SiO_4).

Zn, Fe, Mn—Franklinite ($(Zn, Fe, Mn)_3O_4$).

Zn, S—Sphalerite (ZnS).

Zn, H_2O—Hydrozincite ($3ZnCO_3 + 2H_2O$).

Zn, H_2O—Calamine ($Zn_2SiO_4 + H_2O$).

Water.

H_2O, Al—Bauxite ($Al_2O_3 + 2H_2O$).

H_2O, Al—Kaolin ($Al_2Si_2O_7 + 2H_2O$).

H_2O, Al, P—Turquois($AlPO_4.Al(OH)_3 + H_2O +$ Cu).

H_2O,Al,S—Alunite($K_2SO_4.3Al_2O_3.3SO_3.6H_2O$).

H_2O, As, Cu—Olivenite ($Cu_4As_2O_9 + H_2O$).

H_2O, As, Cu—Conichalcite ($(Cu, Ca)_4As_2O_9 + 3/2H_2O$).

H_2O, Bi—Bismutite ($Bi_2CO_5 + H_2O$).

H_2O, B—Colemanite ($Ca_2B_6O_{11} + 5H_2O$).

H_2O, B, Na—Borax ($Na_2B_4O_7 + 10H_2O$).

H_2O, Ca, S—Gypsum ($CaSO_4 + 2H_2O$).

H_2O, Cu—Malachite ($Cu_2CO_4 + H_2O$).

H_2O, Cu—Azurite ($Cu_3C_2O_7 + H_2O$).

H_2O, Cu—Crysocolla ($CuSiO_3 + 2H_2O$).

H_2O, Fe—Limonite ($2Fe_2O_3 + 3H_2O$).

H_2O, Fe, P—Vivianite ($Fe_3P_2O_8 + 8H_2O$).

H_2O, Mg—Talc ($Mg_3Si_4O_{11} + H_2O$).

H_2O, Mn—Manganite ($Mn_2O_3 + H_2O$).

H_2O, Mn—Psilomelane ($MnO_2 + 2H_2O$).

H_2O, Ni—Garnierite ($H_2(Ni, Mg)SiO_4 + H_2O$).

H_2O, Na—Natron ($Na_2CO_3 + 10H_2O$).

H_2O, Na—Trona ($Na_2CO_3.HNaCO_3 + 2H_2O$).

H_2O, Zn—Hydrozincite ($3ZnCO_3 + 2H_2O$).
H_2O, Zn—Calamine ($Zn_2SiO_4 + H_2O$).

Silicates.

The two mentioned below are the only ones in the above list which will give the test for a silicate described previously.

Willemite (Zn_2SiO_4).
Calamine ($Zn_2SiO_4 + H_2O$).

Carbonates.

Ba—Witherite ($BaCO_3$).
Bi—Bismutite ($Bi_2CO_5 + H_2O$).
Ca—Calcite ($CaCO_3$).
Ca, Mg—Dolomite ($(Ca, Mg)CO_3$).
Cu, H_2O—Malachite ($Cu_2CO_4 + H_2O$).
Cu, H_2O—Azurite ($Cu_3C_2O_7 + H_2O$).
Fe—Siderite ($FeCO_3$).
Pb—Cerussite ($PbCO_3$).
Mg—Magnesite ($MgCO_3$).
Mn—Rhodochrosite ($MnCO_3$).
Na, H_2O—Natron ($Na_2CO_3 + 10H_2O$).
Na, H_2O—Trona ($Na_2CO_3.HNaCO_3 + 2H_2O$).
Sr—Strontianite ($SrCO_3$).
Zn—Smithsonite ($ZnCO_3$).
Zn, H_2O—Hydrozincite ($3ZnCO_3 + 2H_2O$).

CHAPTER VI

THE ELEMENTARY PRINCIPLES OF CHEMISTRY

THE USE AND INTERPRETATION OF CHEMICAL SYMBOLS

Elements. Every body in nature is composed of one of more constituent substances called elements. Sometimes, as in the case of the metals gold, silver, and copper, there is only one substance; it is itself an element. In other cases there are two or more elemental constituents present in the body, which by proper manipulation may be broken up or resolved into its elements.

An element, is then, something that has resisted all attempts to subdivide it into other substances. It follows, necessarily, that an element cannot be formed by a union of other substances.

Each element differs more or less from all others in appearance, properties, and uses. Some are gases, some are opaque and reflect light from the surface—are metals, and some are transparent or translucent—are non-metals. The distinction between metals and non-metals is, however, not sharply marked, since there are elements with intermediate properties. A full list of all known elements is given in the table at the end of this chapter.

It has happened occasionally that a substance supposed to be an element has been found to be composed of two or more elements, and this will probably occur in the future, but chemists feel practically certain of the elementary condition of all the commoner elements. They know now that the Alchemists' search for a method of making gold was foredoomed to failure.

Chemical Compounds. Elements have the property of uniting under certain conditions to form new substances, differing in nature from any of the constituent elements. The results of such unions are not merely mechanical mixtures of the elements in which each component can be identified under the high-power microscope, but are homogeneous substances of definite properties, which will often fail to respond to tests yielded by their constituent elements. Such combinations of two or more elements are called chemical compounds. Thus, the common chemical compound water is composed of one gas, hydrogen, which burns in the air, and another, oxygen, which is essential for respiration, properties quite foreign to water.

Most natural bodies are either chemical compounds or mixtures of them. Other peculiarities possessed by them will be mentioned later.

Alloys. Alloys are rather indefinite compounds

of metals whose natures are not thoroughly understood.

Atoms. Atoms are the smallest particles into which it is believed an element may be divided and still retain all its distinguishing properties. Although not infinitely small, they are far too minute to be seen with the microscope, yet there are abundant more or less indirect proofs of the correctness of the atomic theory. That atoms are themselves subdivisible is now admitted, but these lesser particles reveal the characteristic properties of entirely different elements from those which they formed before disintegration, and are produced only under very unusual conditions, probably never, or very rarely, duplicated in chemical operations.

According to the atomic theory, the atoms of any element have the same weight and size and are identically alike, while the atoms of different elements have different weights and sizes, and further differ to the same extent as do the elements themselves.

Molecules. Molecules are the smallest particles into which it is believed a chemical *compound* may be divided and still remain the same chemical compound. They must consist, evidently, of at least two, and often several, atoms. The same word (molecule) is also applied to the smallest volumes of a gas, even though it be an element instead of a compound, since it appears certain that the smallest particles of gases always consist of at least two atoms.

With certain modifications that it is not necessary to discuss, it may be stated that the number of atoms present in a molecule of a given chemical compound is always the same, and that the different atoms forming a molecule of a certain chemical compound are always present in a fixed proportion. Thus, a molecule of water always contains two atoms of hydrogen and one atom of oxygen.

Symbols. For convenience, the elements are represented by the initial letter, either alone or with an added letter, of their Latin or Greek names, which are in many cases very similar to their English names. These letters are called the symbols of the elements.

Formulæ. The formula of a chemical compound is written by placing the symbols of its component elements in a line, and, if more than one atom of any of these is present in the molecule of the compound, the number of such atoms is indicated by subscripts written after the symbols of the elements thus affected. Thus, the formula of water is H_2O, indicating that the molecule of water contains two atoms of H combined with one atom of O. This formula is read h-two-o.

It is sometimes possible to group the atoms in the formula of a complex substance in such a fashion as to form two or more groups of molecules, indicating that the substance may be formed not only by a union of atoms but also by a combination of molecules. Such formulæ may be written with a period separating the constituent molecules. Thus, $CaCO_3$ (read c-a-c-o-three) is the symbol of calcite, of which marble is a variety, and this formula may be written $CaO.CO_2$, indicating that the material is formed by the union of one molecule of lime (CaO) and one of carbonic acid gas (CO_2). When it is possible to break up a formula into molecules in this way, it is often found that more than one of a certain constituent molecule is present in substance. Thus, the ordinary formula of orthoclase feldspar is $KAlSi_3O_8$, but, after multiplying each atom by two, this is found to consist of one molecule of potassium oxide (K_2O), one of alumina (Al_2O_3), and six of silica (SiO_2), and may therefore be written $K_2O.Al_2O_3.6SiO_2$. The order in which the different molecules are written is of no great importance; the above formula might with equal correctness be written $6SiO_2.K_2O.Al_2O_3$, although this is not the customary order. It should be noticed that a figure prefixed to a molecule, as in the case of the 6 in the above formula, applies only to the *molecule* to which it is prefixed.

Sometimes formulæ like the following (the formula of emerald) are used: $Be_3Al_2(SiO_3)_6$. This is read b-e-three-a-l-two-parenthesis-s-i-o-three-taken-six-times. Both the Si and the O in the parenthesis are affected by the subscript 6 and might be written Si_6O_{18}, but it is sometimes desirable to group elements in parentheses in this way. Groups like this are not molecules since they do not occur as known chemical compounds.

A different use of the parenthesis is seen in the case of those compounds in which the relative proportions of certain compounds are apt to vary—apparently a modification or violation of a previously expressed law. Thus, a common constituent of limestone is a mineral called dolomite, whose formula is $(Mg,Ca)CO_3$. Here the comma between the Mg and Ca indicates that the relative proportion of Mg and Ca is not fixed; there may be a nearly or quite equal number of atoms of both present, or either may predominate to a small or great extent over the other. In an instance like this the element first written in the parenthesis is apt to be the more plentiful. Sometimes it is more convenient (as when the horizontal space is limited) to write the variable elements in such formulæ in a vertical column without using the comma, in this manner: $\left(\begin{matrix}Mg\\Ca\end{matrix}\right)CO_3$.

Atomic weights. The atomic weight of an element is the relative weight of an atom of the element compared with the weight of an atom of H, which is taken as unity, it being the lightest known element. Thus, an atom of Fe is fifty-six times as heavy as an atom of H, so the atomic weight of Fe is 56. A presentation of the methods by which the atomic weights of the various elements are determined is not necessary, but these atomic weights have a practical use which is important. This may be illustrated as follows: Since pure water is composed entirely of molecules having the formula H_2O, if we know the relative weights of the H and O atoms, it should be a simple matter to calculate the proportions by weight of H and O in the molecule, and thus to determine the proportion by weight of these elements in any amount of the substance. There being two atoms of H present in the molecule, each of which weighs one unit (atomic weight of H is 1), and one atom of O, which weighs sixteen units (atomic weight of O is 16), the whole molecule must weigh $2+16=18$ units. It is plain that the H must constitute $^2/_{18}$, or $^1/_9$, and the O form $^{16}/_{18}$, or $^8/_9$ of the whole molecule. It follows that pure water in any amount is $^1/_9$ H and $^8/_9$ O.

In this way it is always possible to calculate the relative proportions of the different elements in a substance whose formula is known, provided no elements involved occur in variable quantities, indicated by placing them, separated by commas, in parentheses, or in vertical columns in parentheses, as previously explained. The atomic weights of all the elements are given in the table at the end of this chapter.

Suppose, for illustration, it be required to find how many pounds of each of the component elements there are in 100 pounds of pure marble, with the formula $CaCO_3$. Let the abbreviation A.W. mean atomic weight, then—

A.W. Ca = 40. Total weight of the one Ca atom = 40 H units.
A.W. C = 12. Total weight of the one C atom = 12 H units.
A.W. O = 16. Total weight of the three O atoms = 48 (3×16) H units.
Total weight of the molecule = 100 H units.
Ca present is $^{40}/_{100}$ of whole = 40% = 40 pounds.
C present is $^{12}/_{100}$ of whole = 12% = 12 pounds.
O present is $^{48}/_{100}$ of whole = 48% = 48 pounds.

As a still more complex case, let it be required to ascertain the amount of Zn and H_2O in 100 pounds of hydrozincite, with the formula $3ZnCO_3 + 2H_2O$.

A.W. Zn = 65.4. Number of atoms of Zn = 3. Weight of three Zn atoms (3×65.4) = 196.2.
A.W. C = 12. Number of atoms of C = 3. Weight of three C atoms (3×12) = 36.
A.W. O = 16. Number of atoms of O = 11. Weight of eleven O atoms (11×16) = 176.
A.W. H = 1. Number of atoms of H = 4. Weight of four H atoms (4×1) = 4.

Weight of molecule (in H units) = 412.2.
Zn present is $^{196.2}/_{412.2}$ of whole = 47.6% = 47.6 pounds.
H_2O present is $^{36}/_{412.2}$ of whole = 8.7% = 8.7 pounds.

Molecular Weights. The molecular weight of a molecule or chemical compound is the sum of the atomic weights of the elements forming the molecule or compound, taking the atomic weight of each element as many times as there are atoms of that element present.

Acids. Acids are substances that impart a red coloration to blue litmus paper. Two kinds are recognized: the halogen acids, compounds of H and one of the so-called halogen elements, Cl, F, Br, and I, and the oxygen acids, compounds of O and H with some other element.

Sometimes several acids are formed by using different proportions of the same elements. The termination "ous" is then used in the name of the one with a relatively low percentage of O, and the termination "ic" in the name of the one with a relatively high percentage of that element. Thus, H_2SO_3 is sulphurous, and H_2SO_4 is sulphuric acid.

Among the other commoner inorganic acids are the following: Hydrochloric (HCl), hydrofluoric (HF), silicic (H_4SiO_4), carbonic (H_2CO_3), nitric (HNO_3), boric or boracic (H_3BO_3), and phosphoric (H_3PO_4).

Bases, Salts, and Chemical Equations. Bases are such substances as impart a blue color to red litmus paper, or which are capable of replacing all or part of the H in an acid and forming a substance called a salt. Bases are always metals, combinations of metal and O, or the latter with the addition of H. Thus, when Na (a base) is added to HCl (an acid), NaCl (a salt) is formed and H is liberated. Such changes as this, called chemical reactions, may be expressed in the form of an equation, thus:

$$HCl + Na = NaCl + H.$$

$$ZnO \text{ (base)} + 2HNO_3 \text{ (acid)} = Zn(NO_3)_2 \text{ (salt)} + H_2O \text{ (water)}.$$

The opposite sides of chemical equations must always contain the same number of atoms of each element involved. When this is true, they are said to balance.

From the above discussion, it is plain that a salt may be defined as a compound that is formed by the reaction of an acid and a base; all or part of the H in the acid is replaced by one or more metals.

If a salt is formed from an acid with the "ous" termination, its name terminates in "ite." Thus, Na_2SO_3 (formed from sulphurous acid) is known as sodium sulph*ite* or sulph*ite* of sodium. When the salt is formed from an acid with the "ic" termination, its name terminates in "ate." Thus, the $Zn(NO_3)_2$ obtained in the reaction recently equated is called zinc nitr*ate* or nitr*ate* of zinc.

In cases where lack of knowledge makes it impossible to decide whether an "ic" or an "ous" acid was involved in the formation of a salt, it is usually safe to use the "ate"—the commoner—termination, calling the salts sulphates, carbonates, phosphates, etc.

Additional Nomenclature. When an element is combined with a non-metal, the resulting compound is named by adding the termination "ide" to the root of the name of the non-metal. The following list will illustrate this usage: $FeAs_2$ is iron arsen*ide* or arsen*ide* of iron, $AgCl_2$ is chlor*ide* of silver, $AuTe_2$ is tellur*ide* of gold, and Fe_2O_3 is ox*ide* of iron.

Sometimes compounds of this type are formed in which more than one element combines with the non-metal, or one element may combine with two different non-metals, as in the following instances: $CuFeS_2$ is copper-iron sulphide, and $Co_2As_2S_2$ ($CoS_2.CoAs_2$) is cobalt arsenide and sulphide. Care should be taken not to confuse the "ite" and "ide" terminations.

TABLE OF ELEMENTS WITH THEIR SYMBOLS AND ATOMIC WEIGHTS

Name.	Symbol.	At. Wts.	Name.	Symbol.	At. Wts.
Aluminum	Al	27.1	Neodymium	Nd	143.6
Antimony	Sb	120.2	Neon	Ne	20
Argon	A	39.9	Nickel	Ni	58.7
Arsenic	As	75	Nitrogen	N	14.04
Barium	Ba	137.4	Osmium	Os	191
Beryllium	Be	9.1	Oxygen	O	16
Bismuth	Bi	208.5	Palladium	Pd	106.5
Boron	B	11	Phosphorus	P	31
Bromine	Br	79.96	Platinum	Pt	194.8
Cadmium	Cd	112.4	Potassium	K	39.15
Caesium	Cs	132.9	Praseodymium	Pr	140.5
Calcium	Ca	40.1	Radium	Ra	225
Carbon	C	12	Rhodium	Rh	103
Cerium	Ce	140.25	Rubidium	Rb	85.4
Chlorine	Cl	35.45	Ruthenium	Ru	101.7
Chromium	Cr	52.1	Samarium	Sm	150
Cobalt	Co	59	Scandium	Sc	44.1
Columbium	Cb	94	Selenium	Se	79.2
Copper	Cu	63.6	Silicon	Si	28.4
Erbium	Er	166	Silver	Ag	107.93
Fluorine	F	19	Sodium	Na	23.05
Gadolinium	Gd	156	Strontium	Sr	87.6
Gallium	Ga	70	Sulphur	S	32.06
Germanium	Ge	72.5	Tantalum	Ta	183
Gold	Au	197.2	Tellurium	Te	127.6
Helium	He	4	Terbium	Tb	160
Hydrogen	H	1.008	Thallium	Tl	204.1
Indium	In	114	Thorium	Th	232.5
Iodine	I	126.85	Thulium	Tm	171
Iridium	Ir	193	Tin	Sn	119
Iron	Fe	55.9	Titanium	Ti	48.1
Krypton	Kr	81.8	Tungsten	W	184
Lanthanum	La	138.9	Uranium	U	238.5
Lead	Pb	206.9	Vanadium	V	51.2
Lithium	Li	7.03	Xenon	Xe	128
Magnesium	Mg	24.36	Ytterbium	Yb	173
Manganese	Mn	55	Yttrium	Yt	89
Mercury	Hg	200	Zinc	Zn	65.4
Molybdenum	Mo	96	Zirconium	Zr	90.6

INDEX

Spotted something that isn't quite right?

Let us know and we'll fix it for future reprints.
And as a thank you for your diligence, we'll
send you a free Sothis book of your choice.

publisher@sothis.press

This page intentionally left blank

Except of course for that line above. And this one.
And whilst we have your attention, perhaps you'd
like to follow us on Instagram @sothis.press. Go on.

Made in the USA
Monee, IL
04 May 2026

49461123R20055